Sit Down, Little

A math story about Carl Friedrich Gauss

by
Jean Colebank

illustrated by
Margeaux Lucas

Saddle Point
PUBLISHING

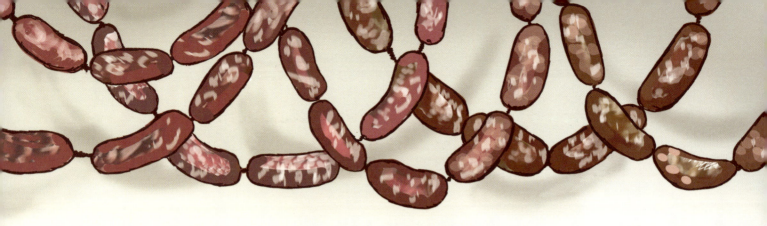

Text copyright ©2019 by Jean Colebank
Illustrations copyright ©2019 by Margeaux Lucas
All rights reserved, including the right of reproduction in whole or in part in any form.
Published by Saddle Point Publishing
P.O. Box 549, Dobbs Ferry, NY 10522
Printed in the USA
ISBN 978-0-9800481-2-4

For Chedda, friend extraordinaire — ML

To Mom and Dad: for all the words, music, and *numbers* you brought to my life — JC

Many years ago in Germany, there was a little boy with a very long name: Johann Carl Friedrich Gauss. He was the only child of hardworking parents who treated him royally.

His mother made him
velvet coats,
satin britches,
and silk shirts
with ruffles.

His father made him
soft leather shoes.

Soon everyone called him Little Prince. They knew he was spoiled. What they didn't know was that Johann had a secret. His head was full of numbers and he couldn't keep still.

One morning, he stopped at the meat market.

"You have 38 sausages hanging from the ceiling," he told the butcher.

"Don't bother me!" growled the butcher. "I'm too busy for you!"

He walked by the bakery and heard someone shouting, "Where are my glasses?"

Little Prince popped in. "I can read the recipes for you," he announced.

"Go away," huffed the baker. "You can't read— you're too young!"

Johann ran to the park to watch a stickball game.

Peering over the shoulder of the scorekeeper, he said, "Your totals are incorrect."

"Get lost," said the scorekeeper. "I don't need you around!"

Disappointed, Johann hung his head and walked home.

He sat down next to his father, who was checking over the household bills. "Look again, Papa," said Johann. "You've made some mistakes."

His father checked again. "Why, you're right!" he exclaimed. "Wife, our child can add and subtract! He must go to school!"

"Oh my," sighed Johann's mother. "School costs a lot of money. How will we ever afford that?"

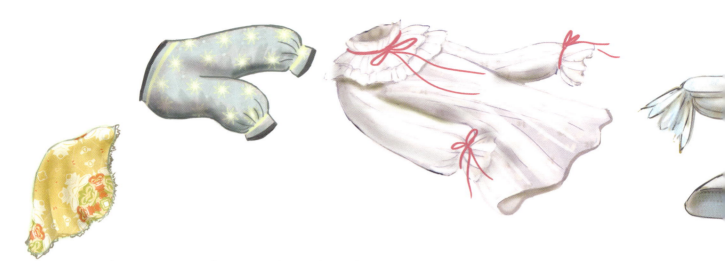

Johann ran to his trunk and took out his silk shirts and satin britches.

He piled up bars of soap his mother had made.

He rolled out a barrel of sauerkraut stored on the back porch.

"Sell these!" he cried. "I want to go to school!"

Johann's mother and father went to see Schoolmaster Buttner, who was not eager to have an additional student in his class.

"The students I have are already enough to control," complained the teacher. "Why should I take your son?"

"Our son is a genius," said Johann's father. "He can read and write and knows arithmetic."

The schoolmaster frowned. "What is your son's name?"

"His name is Johann, but we call him Little Prince," answered Johann's mother.

"Very well. I shall give this prince of yours one week to prove himself worthy to be in my class."

On his first day of school, Little Prince sat quietly—until he saw something green and squirmy crawling out of the teacher's desk.

"Sir! Sir!" he cried, jumping up and pointing at the critter.

"WHO DID THIS?"

the schoolmaster bellowed.

No one said a word.

During morning recess,
Little Prince joined the children outside.

"Catch this," a boy said as he thrust the ball quickly at Johann.

"Tag me if you can reach me!" called another boy, running fast up
a steep hill. Johann sped up and quickly tagged the boy on the shoulder.

"Morning recess is going by too fast," thought Little Prince.
"I'm having so much fun!"

A tall boy took Little Prince aside. "I have some advice for you,"
he whispered. "Don't give Schoolmaster Buttner any correct answers.
If you know too much, he'll want you to be his assistant.
He'll keep you inside at recess time to correct slates
and you won't be able to play with us."

Johann remembered the advice. When the teacher asked him to read the 1 to 100 chart, he stood up and purposely mixed up the 30s and the 60s. "Sit down, Little Prince," said the teacher.

When Mr. Buttner asked him, "How much is 5 + 6?" he stood up and replied, "12."

"Sit down, Little Prince!" snapped Mr. Buttner. "Study the addition problems some more."

Little Prince was growing restless.

School was not as exciting as he thought it would be.

Recess was fun, but class time was not.

It didn't matter if he gave wrong answers; he knew the correct ones, and he wasn't being asked anything new.

Little Prince grew more restless each day.

He tapped his chalk on his slate.

He snapped his fingers.

He clicked his heels together.

Finally, one afternoon he called out, "I'm bored!"

Everyone gasped.

No one had ever spoken like that before in class.

"Aha," said Schoolmaster Buttner. "You want something to do? I'll give you something to do. Add up all the numbers from 1 to 100—and the answer must be correct! Use your chalk and slate to figure it out. I'll give you an hour."

Little Prince thought a minute and smiled.

"I don't need my chalk and slate or an hour. I know the answer. It's 5,050."

Schoolmaster Buttner was stunned. He could hardly speak.

"How...how did you calculate that so quickly?" he stuttered.

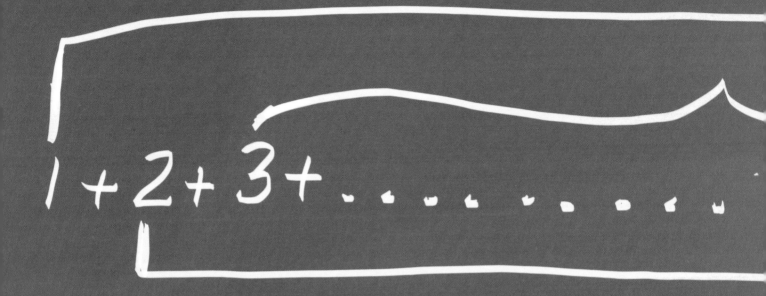

"I just thought, hmmm—
1 + 100 = 101,
2 + 99 = 101,
3 + 98 = 101,
and so on,

and I realized in a flash that every pair
would give me a sum of 101.

There are 50 pairs, so 50 × 101
will give me the answer.

Since 50 × 100 = 5,000
and 50 × 1 = 50,
the answer is 5,000 + 50 or 5,050."

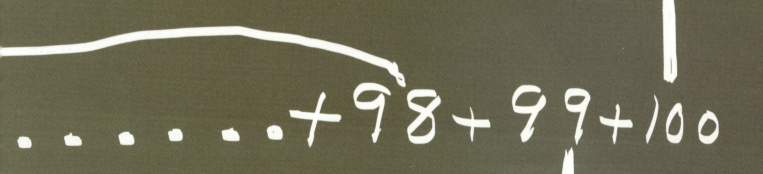

Little Prince sat down.

The class became quiet.

From then on, everyone in class wanted to add up all kinds of numbers.

They had no time to think about slipping little critters into Mr. Buttner's desk.

When the week was over,
Schoolmaster Buttner met with Johann's parents.

"I can't have Little Prince in my class," he said.

"He needs to go to the university.
Your son is a genius, and surely a mathematical one!"

And that's exactly what Little Prince did.

He studied very hard and became—

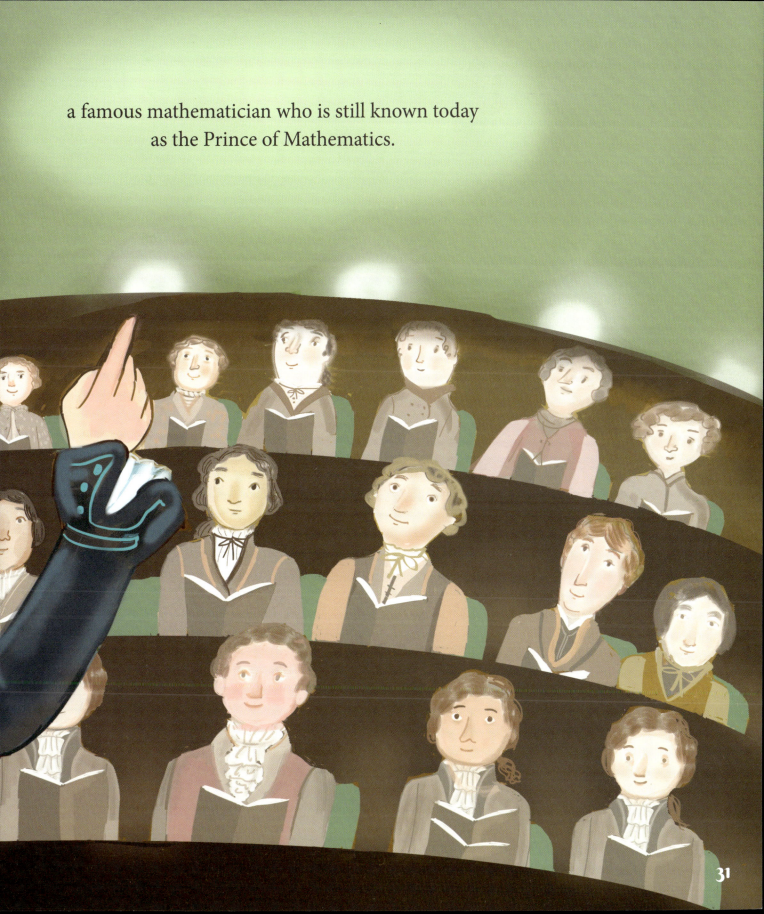

a famous mathematician who is still known today as the Prince of Mathematics.

JOHANN CARL FRIEDRICH GAUSS
1777–1855

NOTES FROM THE AUTHOR

This story is fictional, but many of the events are real. There really was a mathematician named Johann Carl Friedrich Gauss.

He was born in Brunswick, Germany. In elementary school, he did have a teacher named Büttner. As a young boy, Gauss figured out a method to quickly add a long list of numbers. Below is a formula that describes his steps.

$$S_n = \left(\frac{n}{2}\right) \times (a_1 + a_n)$$

$$Sum = \left(\frac{\text{number of numbers}}{2}\right) \times (\text{first number} + \text{last number})$$

$$Sum = \left(\frac{100}{2}\right) \times (1 + 100)$$
$$= 50 \times 101$$
$$= 5{,}050$$

The formula will work on any set of numbers as long as the same number is added from one number to the next. In Example A, seven is added to each number to get to the next number, and in Example B, four is added each time.

Example A:

$3 + 10 + 17 + 24$

$$Sum = \left(\frac{4}{2}\right) \times (3 + 24)$$
$$= 2 \times 27$$
$$= 54$$

Example B:

$4 + 8 + 12 + 16 + 20$

$$Sum = \left(\frac{5}{2}\right) \times (4 + 20)$$
$$= 2\tfrac{1}{2} \times 24$$
$$= 60$$

Notice that a set can have an odd number of numbers. If so, it will include half a pair!

Gauss is one of the most accomplished and influential mathematicians of all time. One of his mottos was "Few, but ripe." This means that he worked on lots of mathematical ideas but published them only when he was sure they were absolutely complete and correct.

Gauss made discoveries in many areas of math and science, including algebra, electrostatics, geophysics, number theory, optics, statistics, and more. He earned the title and is still known to this day as the Prince of Mathematics.